PROCEEDINGS

OF THE

CAMBRIDGE PHILOSOPHICAL SOCIETY

(SUPPLEMENT)

NIELS BOHR

ON THE APPLICATION OF THE QUANTUM
THEORY TO ATOMIC STRUCTURE

PART I

THE FUNDAMENTAL POSTULATES

CAMBRIDGE

AT THE UNIVERSITY PRESS

1924

CAMBRIDGE UNIVERSITY PRESS
Cambridge, New York, Melbourne, Madrid, Cape Town,
Singapore, São Paulo, Delhi, Tokyo, Mexico City

Cambridge University Press
The Edinburgh Building, Cambridge CB2 8RU, UK

Published in the United States of America by Cambridge University Press, New York

www.cambridge.org
Information on this title: www.cambridge.org/9781107681583

First published 1924
First paperback edition 2011

A catalogue record for this publication is available from the British Library

ISBN 978-1-107-68158-3 Paperback

ON THE APPLICATION OF THE QUANTUM THEORY TO ATOMIC STRUCTURE*

By NIELS BOHR.

From the *Zeitschrift für Physik*, **13**, p. 117, 1923, with the consent of the Editor. Translated by L. F. Curtiss, National Research Fellow (U.S.A.), with the author's concurrence. Communicated to the Cambridge Philosophical Society by Mr R. H. Fowler, Trinity College.

I. THE FUNDAMENTAL POSTULATES OF THE QUANTUM THEORY

INTRODUCTION

The quantum theory, which can be regarded as a rational generalisation of the original assumptions of Planck concerning the exchange of energy between a simple harmonic oscillator and a field of radiation, presents a sharp departure from the ideas of classical electrodynamics in the introduction of discontinuities into the laws of nature. From the present point of view of physics, however, every description of natural processes must be based on ideas which have been introduced and defined by the classical theory. The question therefore arises, whether it is possible to present the principles of the quantum theory in such a way that their application appears free from contradiction. It is our purpose to investigate this question more closely in this essay. We shall show how, by the development of the quantum theory, a foundation seems to be created for a theory which accounts for numerous phenomena for which the classical theory obviously fails, and at the same time closely follows in a natural manner the applications in which the classical theory has given such good service. In this connection we must always keep clearly before us the far-reaching character of the departure from our customary ideas, which is effected by the introduction of discontinuities. This is very important when

---

* The present communication forms the first of a series of essays which will appear under this title. Their object is to expound systematically the problems which have been met in the investigation of atomic structure. In method of treatment, they follow closely a work published in 1918, among the papers of the Copenhagen Academy ("On the Quantum Theory of Line Spectra," *Det Kgl. Danske Vidensk. Selsk. Skrifter*, 8. Raekke, IV, 1), of which a German translation has recently appeared from the press of Fried. Vieweg & Son, A.-G., Braunschweig. In the future this work will be cited as *Q. o. L.* In substance, they follow closely a summarised form of this work, which appeared as a translation of a Copenhagen lecture in *Zs. für Phys.* 9, 1, 1922. Together with two earlier lectures, this has been published as a separate volume of Vieweg's series under the title *Drei Aufsätze über Spektren und Atombau*. [The three essays—*The Theory of Spectra and Atomic Constitution*, Camb. Univ. Press, 1922—are substantially the English equivalent.]

dealing with the question of the more exact formulation of the principles of the quantum theory, in connection with which, in the present state of knowledge, one is forced always to keep in mind the range of applicability of the theory. In order to make our task as simple as possible, we shall choose as a starting point a formulation which has proved itself suitable, in a limited region, to the treatment of a few typical applications for which the classical theory was particularly helpless. We begin by considering an isolated atomic system, by which we understand a system of electrically charged particles, moving in such a way under the influence of their mutual forces that the distance between the particles always remains below a certain definite limit. The application of the quantum theory to such a system is characterised in its main features by two fundamental postulates which we wish especially to discuss and explain. We shall also find here in passing an opportunity to go into the range of applicability of these postulates, and their possible modification for non-isolated systems.

CHAPTER I

THE STATIONARY STATES

§ 1. *The First Fundamental Postulate.* The first postulate of the quantum theory for an isolated atomic system states that, among the kinematically conceivable relative motions of the particles of the atom, there exist certain states, the so-called *stationary states*, which are distinguished by a peculiar *stability*, shown by the fact that every permanent change in the motion of the isolated system must consist in a complete transition from the original to another of these stationary states.

The contradiction between the demands contained in this postulate and the classical theory is illuminating. According to the latter theory a system of electrically charged particles not only will have no particular motions which, in contrast with the others, exhibit a stability of the kind discussed, but will possess no fixed motions at all, since every motion of the particles relative to each other, which can be conceived, will give rise to the emission of electromagnetic radiation. If it has, nevertheless, been possible in the description of the motion in the stationary states to use to a great extent conceptions obtained from the classical theory, this comes foremost from the fact that, in the atomic systems ordinarily considered, the change in the motion of the particles which according to this theory would be directly connected with the emission of radiation, is at any moment only small in comparison with the change of the motion which arises from the electromagnetic forces acting between the particles. These forces may arise from the mutual attractions and repulsions of the particles, or from magnetic fields which originate in the motion of the particles. The attempt to keep a connection with classical ideas leads therefore immediately to the assumption that the motion within the stationary states can be described, to a close approximation, by the laws of ordinary electrodynamics, if only one neglects the reaction connected with the emission of radiation. This assumption means that the motion of the system in the stationary states is governed very approximately by laws which are stated in their clearest form in the so-called canonical equations of motion:

$$\frac{dp_k}{dt} = -\frac{\partial E}{\partial q_k}, \quad \frac{dq_k}{dt} = \frac{\partial E}{\partial p_k} \quad (k = 1, ..., s). \quad ......(1)$$

s is the number of degrees of freedom, by which we mean the number of independent variables which are necessary for the description of the positions of the particles, relative to a frame of reference in which the system as a whole may be regarded as at rest. $q_1, ..., q_s$ are a set of generalised space-coordinates which determine the positions of the particles relative to such a frame of reference, while $p_1, ..., p_s$ are the momenta conjugated to them. E is a function of the p's and q's, which can be regarded as the total energy of the system. It is defined to a close approximation, according to the classical theory, by the relative positions and velocities of the particles.

The constancy of the energy during the motion in the stationary states immediately follows from equations (1). Apart from this, the solution of equations (1) is, in general, of a very complicated character and scarcely offers sufficient basis for fixing and describing discrete stationary states of the system. For this purpose certain pronounced *periodic properties* in the motion seem to be necessary*. In the cases where it has been possible to reach a rational method of fixing the stationary states on the basis of the equations of motion (1), their general solution has in fact been of a so-called simply or multiply periodic character. In more complicated cases we must be prepared for the fact that the equations (1) will not prove sufficient for the description of the motion in the stationary states even to the degree of approximation conditioned by the radiation-reaction, which we mentioned above (cf. p. 15).

* In Planck's general treatment of the application of the quantum theory to atomic systems (*Berl. Ber.* 1918, p. 435), which, however, is based on physical points of view differing essentially from our fundamental postulates, it is regarded as a necessary condition for a quantisation, that, in systems of more than one degree of freedom, in addition to the energy integral at least $s-1$ other uniform integrals of equations (1) exist, which can serve to define definite s-dimensional regions of the $2s$-dimensional phase-space, within which the representative point of the orbit remains during the motion. As Kneser (*Math. Ann.* **84**, 277, 1921) has shown, such a demand is essentially equivalent to the presence of periodic properties of the kind mentioned in the general solution of the equations of motion.

§ 2. *The Fixation of Stationary States for Simply and Multiply Periodic Systems.* We define as systems of this kind those for which each motion described by equations (1) is so constituted that the displacement of any individual particle in space, apart from an eventually uniform translation of the system as a whole, can be resolved into a series of harmonic vibrations. The displacement of a particle in a given direction can be expressed as a function of the time in the following manner:

$$\xi = \Sigma C_{\tau_1 \ldots \tau_u} \cos 2\pi \left([\tau_1 \omega_1 + \ldots + \tau_u \omega_u] t + \gamma_{\tau_1 \ldots \tau_u} \right). \quad \ldots\ldots(2)$$

$\omega_1, \ldots, \omega_u$ are the so-called fundamental frequencies, whose number, u, we will call the "degree of periodicity." The summation is to be extended to all integral values of the numbers $\tau_1, \ldots, \tau_u$. The uniqueness of the solution is conditioned by the fact that among the quantities $\omega_1, \ldots, \omega_u$ there exist no relations of the form:

$$m_1 \omega_1 + \ldots + m_u \omega_u = 0, \quad \ldots\ldots(3)$$

where $m_1, \ldots, m_u$ are a series of whole numbers.

The stationary states of such a system are fixed by a number of conditions which can be regarded as rational generalisations of the original assumptions of Planck concerning the selected states of a simple harmonic oscillator. These "conditions of state" (Zustandsbedingungen), the number of which is equal to the degree of periodicity, can be written in the following form:

$$J_1 = n_1 h, \ldots, J_u = n_u h, \quad \ldots\ldots(A)$$

where h is Planck's constant, and $n_1, \ldots, n_u$ a series of whole numbers, the so-called "quantum numbers", while $J_1, \ldots, J_u$ are certain quantities determining the motion of the system. They are closely related to the periodic properties of the system, and are most simply defined as the conjugated momenta of a number of analytic variables, which can be suitably designated as "uniformising variables". These variables may be characterised in the following manner:

The generalised coordinates $q_1, \ldots, q_s$ entering into (1), and their conjugated momenta $p_1, \ldots, p_s$, can be expressed in terms of the following new system of s pairs of canonically conjugated variables:

$$\begin{aligned} w_1, w_2, \ldots, w_u; & \quad \beta_1, \ldots, \beta_{s-u}, \\ J_1, J_2, \ldots, J_u; & \quad \alpha_1, \ldots, \alpha_{s-u}, \end{aligned} \quad \ldots\ldots(4)$$

where the first series of variables is to be substituted for the q's, the second for the p's, in the canonical equations (1). These new variables are to satisfy the following conditions:

I. The q's and p's are periodic in each of the variables $w_1, \ldots, w_u$ with the period unity; that is, each coordinate q_r can be written as an infinite, multiple trigonometric series of the type:

$$q_r = \Sigma C_{\tau_1 \ldots \tau_u} \cos 2\pi \left(\tau_1 w_1 + \ldots + \tau_u w_u + \gamma_{\tau_1 \ldots \tau_u} \right), \quad \ldots(5)$$

where the quantities C and γ depend only on the J's, the α's, and the β's. The summation is to extend to all combinations of integral values of $\tau_1, ..., \tau_u$.

II. The energy of the system, regarded as a function of the new variables, depends only on the quantities $J_1, ..., J_u$. As a consequence of the canonical equations, this condition implies that the variables $J_1, ..., J_u$, as well as the α's and the β's, remain constant during any motion, while the variables $w_1, ..., w_u$ vary uniformly with the time:

$$w_r = \omega_r t + \delta_r \quad (r = 1, ..., u), \qquad \text{......(6)}$$

where
$$\omega_r = \frac{\partial E}{\partial J_r} \quad (r = 1, ..., u). \qquad \text{......(7)}$$

From (6) and from the condition I, it now further follows that every coordinate q_r (and therefore the components of the resulting electric moment of the system in a given direction) can be expressed as a function of the time just by an expression of the form (2).

III. The quantities J_r, in the fixation of which an additive constant has so far remained undetermined, are to be so determined that the integral

$$\int_{t_0}^{t} \Sigma p_r \, dq_r,$$

extended over a mechanical motion of the system, which is the ordinary "quantity of action" independent, as is well known, of the choice of coordinates, differs from the "uniformised" quantity of action,

$$\int_{t_0}^{t} \sum_1^u J_r \, dw_r = (t - t_0) \sum_1^u J_r \omega_r, \qquad \text{......(8)}$$

for each motion of the system only by terms periodic in the time.

The properties of the uniformising variables imply that the assumption, that no linear relations of the type (3) shall exist among the quantities ω_r, does not limit the generality. If such a relation exists, one can always, by the well-known proper transformation, replace the variables J and w by linear combinations of these variables with integral coefficients, so that the number u of the pairs of variables, J_r, w_r, will be reduced by unity, while simultaneously the pairs of conjugated quantities α, β, which we shall often designate as orbital constants, will be increased by unity*.

---

* The development of the conditions for fixing the stationary states has gradually advanced by contributions from a great number of authors, among whom is Planck himself. We shall not go further into this development here, since it is presented in *Q. o. L.* with references to the literature. We shall only recall a few points here concerning more recent work. For a purely periodic system ($u=1$) condition (A) is identical with the statement that the action integral, taken over a period, is equal to a whole multiple of Planck's constant. For multiply periodic systems in which the equations of motion can be solved by "separation of the variables", that is, where it is possible to find a group of space-coordinates $q_1, ..., q_s$, so that each of

§ 3. The Fixation of the Stationary States of a System in Presence of an External Conservative Field of Force. In an attempt to explain more clearly the conditions which fix the stationary states, the first question to arise is: How is the demand of the quantum theory for the stability of the stationary states related to the consequences of the classical theory concerning the effect of external forces on such a system, or the interaction of two such systems? In the discussion of such effects, we shall begin by investigating the case in which the external forces form a conservative field, constant with respect to the time. If, in this case, the solution of the equations of motion —which are given by (1) if the potential energy of the system with reference to the external forces is included in the energy function —again proves to be of a multiply·periodic character, we meet a problem which is not essentially different from that of fixing the stationary states of an isolated system, and we shall assume that the system in the presence of external forces will possess a series of stationary states which are fixed by the conditions (A).

In the case, common in physical applications, in which the external forces are only small compared with the forces acting between the particles, the question of the effect of the external forces can be treated in a more comprehensive manner, since the changes in the stationary states can be brought into immediate relation with the changes of the motion of the system arising from the external forces, the so-called "perturbations". According to the usual procedure of analytical mechanics, these perturbations will

the conjugated momenta p_r depends only on the corresponding space-coordinate q_r during the motion, where further the degree of periodicity of the motion is equal to the number of degrees of freedom ($u = s$), the stationary states are fixed by the conditions (A), (established by Wilson and Sommerfeld, and especially by the theory developed by Epstein), if each of the quantities J_r is set equal to the "separated" elements of action $\int p_r dq_r$, where the integrals extend over a complete period of change of the corresponding q_r. If the degree of periodicity is less than the number of degrees of freedom, the system is often called "degenerate." In this case, the fixation of the quantities J, appearing in the conditions (A), will not be given immediately by a possible solution of the equations of motion by means of separation of the variables. This is explained by the fact that the action elements mentioned above may convey various meanings, since, in such a case, a separation is often possible in different systems of coordinates. Within the general class of multiply periodic systems, the systems which permit a separation of variables form a family, the motions of which can be regarded as a form of transition between the simple case in which the motion resolves into several components purely periodic in time, which correspond to the various "independent degrees of freedom," and the general case of motions which can be resolved according to formula (2) into harmonic components. All known motions of this kind can be described by a set of uniformising canonical variables, as discussed in the text. With reference to the typical applications in celestial mechanics, such variables are frequently called "angle variables." The application of the analytical theory of uniformising variables in the quantum theory began, as is well known, with Schwarzschild. A summarising presentation of the theory is given in a dissertation by Burgers (*Het Atoommodel van Rutherford-Bohr*, Haarlem, 1918). This author has made the theory of the stationary states in multiply periodic systems essentially precise in that, for the first time, he has introduced a condition equivalent to condition III (see note on p. 13).

be described by considering at each moment the so-called "oscu-lating" motion, namely, the motion which would ensue at the moment under consideration if the external forces were suddenly to vanish. Since we assume that the undisturbed motion possesses a multiply periodic character, the osculating motion can be de-scribed by a number of uniformising variables of the kind indicated above. The change of these variables with the time will then be given by the following equations:

$$\frac{dJ_r}{dt} = -\epsilon\,\frac{\partial\Omega}{\partial w_r}; \quad \frac{dw_r}{dt} = \omega_r + \epsilon\,\frac{\partial\Omega}{\partial J_r} \quad (r = 1, \ldots, u),$$

$$\frac{d\alpha_i}{dt} = -\epsilon\,\frac{\partial\Omega}{\partial\beta_i}; \quad \frac{d\beta_i}{dt} = \epsilon\,\frac{\partial\Omega}{\partial\alpha_i} \quad (i = 1, \ldots, s-u), \qquad \ldots\ldots(9)$$

where $\epsilon\Omega$ represents the potential energy of the external forces, regarded as a function of the osculating uniformising variables. The constant factor ϵ is a small quantity proportional to the in-tensity of the external forces. The function Ω, as a result of condition I for the uniformising variables, may be written in the following manner:

$$\Omega = \Psi_0\,(J_1, \ldots, J_u, \alpha_1, \ldots, \alpha_{s-u}, \beta_1, \ldots, \beta_{s-u})$$
$$+ \sum_{\tau_1 \ldots \tau_u} \Psi_{\tau_1 \ldots \tau_u} \cos 2\pi\,\{\tau_1 w_1 + \ldots + \tau_u w_u + \gamma_{\tau_1 \ldots \tau_u}\}, \qquad \ldots\ldots(10)$$

where the quantities Ψ in the second term of the right-hand side, as well as the quantities Ψ_0, depend on the arguments,

$$J_1, \ldots, J_u, \alpha_1, \ldots, \alpha_{s-u}, \beta_1, \ldots, \beta_{s-u}.$$

The same is true for the quantities γ. The summation is to be carried out for all combinations of positive and negative integral values of the τ's, with the exception of the combination

$$\tau_1 = \tau_2 = \ldots = \tau_u = 0.$$

The first term, which corresponds to this combination, is propor-tional to the mean value of the potential of the external forces, taken over a complete motion of the undisturbed system.

The character of the perturbations determined by (9) will be essentially different according as the quantity Ψ_0 depends only on the quantities $J_1, \ldots, J_u$, just as the energy function of the un-disturbed motion, or also contains in addition the quantities α and β, which will be the case in general when u is less than s. In the first case, the perturbations will always possess a multiply periodic character and the stationary states of the disturbed system will be fixed by the same number of conditions as the stationary states of the undisturbed system. In order to specify analytically the explicit conditions for the stationary states of the disturbed system, we must, however, seek new variables which are suitable for uni-formising this system. Such a change of variables will, of course, be a so-called contact transformation. Here the transformation will be especially simple, since, as a result of the assumed smallness of ϵ, the new uniformising variables, which we shall distinguish by a dash, deviate only slightly in their significance from the uniformising variables of the original system. By neglecting higher powers of ϵ, we can write, according to the laws for infinitesimal contact transformations:

$$J_r{'} = J_r + \epsilon \frac{\partial S}{\partial w_r}; \quad \alpha_i{'} = \alpha_i + \epsilon \frac{\partial S}{\partial \beta_i}, \quad (r = 1, ..., u),$$

$$w_r{'} = w_r - \epsilon \frac{\partial S}{\partial J_r}; \quad \beta_i{'} = \beta_i - \epsilon \frac{\partial S}{\partial \alpha_i}, \quad (i = 1, ..., s - u),$$

$$\left. \right\} \quad ...(11)$$

where S is a function of J, w, α, and β. One now easily finds, by placing

$$S = + \frac{1}{2\pi} \sum_{\tau_1...\tau_u} \frac{\Psi'_{\tau_1...\tau_u}}{\tau_1 \omega_1 + ... + \tau_u \omega_u} \sin 2\pi (\tau_1 w_1 + ... + \tau_u w_u + \gamma_{\tau_1...\tau_u}) \quad ...(12)$$

where the summation is to extend over all integral combinations of the τ's, except $\tau_1 = \tau_2 = ... = \tau_u = 0$, that the new variables in fact fulfil conditions I, II, III, which are necessary in order that they may represent uniformising variables for the new system. The total energy for the perturbed system, if we again, as everywhere in the calculation, neglect the second and higher powers of ϵ, is given by the following expression:

$$E' = E (J_1{'}, ..., J_u{'}) + \epsilon \Psi_0 (J_1{'}, ..., J_u{'}), \quad ......(13)$$

where E represents the energy function of the undisturbed system. The stationary states are determined by equations (A) if we only introduce the quantities $J_1{'}, ..., J_u{'}$ in place of $J_1, ..., J_u$:

$$J_k{'} = n_k h \quad (k = 1, ..., u). \quad ......(14)$$

From the expression (13) it immediately follows that the change in the energy of the stationary states as a result of the presence of external forces, to a first approximation, is simply equal to the mean potential energy of the system with reference to the external field, taken over a complete motion of the undisturbed system.

In the case where Ψ_0 depends on the quantities α and β in addition to J', the perturbations will be of an essentially different kind, since the so-called "secular perturbations" will now enter. In additions to oscillations of a multiply periodic character in α and β, with periods equal to those of the undisturbed motion, and with amplitudes proportional to the external forces, these quantities will, in fact, as is clear from (9) and (10), be subjected to slow changes which in the course of time will introduce finite differences in the motion of the system. If these changes are of a simply or multiply periodic character, the motion of the disturbed system will again be multiply periodic, but with a higher degree of periodicity than that of the undisturbed motion. In addition to the frequencies which correspond to the fundamental frequencies of the undisturbed system, there will now as a result of the secular perturbations be a further number of frequencies the magnitudes of which will be proportional to the external forces.

For the analytical treatment of this problem it is not sufficient to carry out a simple contact transformation. If we again carry out the transformation defined by (11) and (12), we do of course make the quantities w vanish from the expression for the energy, but not the quantities α and β. For the energy of the perturbed system we shall, in fact, obtain the form:

$$E' = E\,(J_1', ..., J_u') + \epsilon \Psi_0\,(J_1', ..., J_u', \alpha_1', ..., \alpha'_{s-u}, \beta_1', ..., \beta'_{s-u}),$$
$$\dots\dots(15)$$

where the quantities α' and β', in general, change with the time. By neglecting small quantities proportional to ϵ^2, this change will be given by the equations:

$$\frac{d\alpha_k'}{dt} = -\epsilon\,\frac{\partial \Psi_0}{\partial \beta_k}; \quad \frac{d\beta_k'}{dt} = \epsilon\,\frac{\partial \Psi_0}{\partial \alpha_k} \quad (k = 1, ..., s - u). \dots\dots(16)$$

These equations, which describe the secular perturbations, possess the same canonical form as the equations of motion (1). The problem which we meet in fixing the stationary states by use of the conditions (A) is therefore referred back to a problem which is completely analogous to the fixation of the stationary states of a system of $s - u$ degrees of freedom. In the case in which the solution of equations (16) possesses a simply or multiply periodic character of a degree of periodicity $u' - u$, it will thus be possible to introduce a group of uniformising variables,

$$w'_{u+1}, ..., w'_{u'} \qquad \beta_1'', ..., \beta''_{s-u'}$$
$$J'_{u+1}, ..., J'_{u'} \qquad \alpha_1'', ..., \alpha''_{s-u'}$$

which are suitable for the description of the secular perturbations given by (16), just as the variables (4) serve for the motion of a multiply periodic system represented by the canonical equations (1). The stationary states will now be determined by u' conditions of the type (A), that is, to the u conditions (14), $u' - u$ conditions are to be added:

$$J_l' = n_l h \quad (l = u + 1, u + 2, ..., u'). \qquad \dots\dots(17)$$

The energy of the stationary states is given by

$$E' = E\,(J_1', ..., J_u') + \epsilon \Psi_0'\,(J_1', ..., J_u', J'_{u+1}, ..., J'_{u'}), \quad \dots(18)$$

where the second term of the right-hand side, as in equation (15), represents the mean value of the disturbing potential taken over a motion of the undisturbed system. In this case, of course, in addition to the conditions (A) of the stationary states of an undisturbed system, this motion is subjected to the further conditions which are expressed by (17), in the limiting case in which the external forces vanish ($\epsilon = 0$). Since these conditions, from the nature of the case, depend completely on the periodic properties of the secular perturbations, they will, with different fields of force, determine entirely different orbital properties. Under the assumption that the motion in the stationary states in the presence of external fields of force can be described by ordinary mechanics, it therefore already appears, as a necessary condition with respect to the stability of the stationary states, that the motion of the undisturbed system is determined by a number of conditions, which is not greater than the degree of periodicity.

In connection with the question as to the degree of exactness with which the stationary states of the perturbed system are fixed, it may be recalled that we cannot expect to fix the stationary states of an atomic system with greater exactness than is consistent with the description of the motion by equations (1), the approximate character of which is already conditioned by the neglect of the radiation-reaction appearing in the classical theory. This circumstance is above all others important, if there is a question of following the motion of the system during times which are very long compared with the periods of the undisturbed motion, on the basis of the equations of the theory of perturbations. We shall discuss a problem connected with this question in Chapter II*.

§ 4. *The Stability of the Stationary States in the Presence of External Varying Fields of Force. The Adiabatic Principle.* We now wish to consider an atomic system which is exposed to the influence of

---

* The theory given in this section for a perturbed system is presented in its main features in *Q. o. L.* Part II, § 2, where the treatment is limited to the case, especially important for applications to atomic questions, in which the motion of the undisturbed system is simply periodic. The presentation there given follows closely the physical points of view described in the following sections of this essay as the Adiabatic and the Correspondence Principles. For the simple analytical treatment given here, I owe thanks to the cooperation of Dr Kramers (cf. H. A. Kramers, *Zs. für Phys.* **3**, 199, 1920). See also Burgers' dissertation (cf. footnote, p. 6), where several instructive applications of the methods of the theory of perturbations are given to problems of the quantum theory. In a series of articles which have recently appeared, Epstein has further treated the problem of perturbed systems (*Zs. für Phys.* **8**, 211, 305; **9**, 92, 1922). On several points he takes a point of view which is different from that presented in *Q. o. L.* The difference, however, seems essentially to be based on his method of calculation, which scarcely permits a sufficiently general application to justify the conclusions of this author. In this connection a work by Born and Pauli (*Zs. für Phys.* **10**, 137, 1922) may be mentioned. They present the quantum theory of perturbed systems from the same point of view as in the text above. However, they describe more closely in what way the treatment of the perturbed systems can be carried out with a higher degree of approximation. With reference to this point, the question of the convergence of series of the kind appearing in equation (12) is discussed. In answering this question it may be essential to consider the circumstance last alluded to in the text.

an external field of force, varying with the time. We meet characteristic examples of such cases when an atom is affected by electromagnetic radiation, or when we consider a collision between two
atoms. In such cases, the postulates of the quantum theory concerning the existence and the stability of the stationary states lead
to the following demand: *In a process in which an atom is exposed to
temporary external effects, or in the case of several atomic systems
entering into mutual interaction, each of the atomic systems concerned
will be in stationary states, after, as well as before the process* *. The
result of this demand is that the interaction of an atomic system
with respect to external effects in general cannot, even to a first
approximation, be described by use of the laws of classical electrodynamics. This failure in principle of the electrodynamic laws
might at first sight seem peculiar, in consideration of the fact that
our descriptions of many physical phenomena rest essentially on
an application of these laws to the reaction of atomic systems with
reference to external influences. However, it must be recalled that,
in the case of such phenomena as, for example, those of the kinetic
theory of heat, it is often a question of processes in which the states
of the atomic systems concerned correspond to large quantum
numbers,—if on the whole the postulates of the quantum theory in
their simple form are valid for these (see the following chapter).
Consequently the motions in successive stationary states are
relatively little different, and the exchange of energy in the process
corresponds to transitions in which the quantum numbers change
by many units.

The applicability of the electrodynamic (mechanical) laws in
this limiting region may, therefore, be regarded merely as a con-
cealment of the difference in principle between the laws which
govern the actual mechanism of the process and the continuous
laws of classical conceptions. The task which comes to us at this
point in the development of the quantum theory may, therefore,
be formulated as the seeking out of the quantitative laws for
reactions of atomic systems, which, in the limiting region mentioned,
follow the statistical results obtained through the classical theory,
and at the same time also agree with the characteristic conditions

* See N. Bohr, *Phil. Mag.* 26, 13, 1913. This paper appeared in a German
translation as *Abhandlung über Atombau*, Fried. Vieweg & Son, 1920 (cited in the
future as *Abhl. über A.*). Cf. *Abhl. über A.* I, p. 19. The question of the insufficiency
of the classical theory for describing the interaction of an atom with a field of
radiation is more exactly discussed in Chapters II and III of this essay. A survey
of the experimental results of the collisions between free electrons and atoms, for
which primarily we must thank J. Franck and his co-workers, is given by this author
in two summarising essays, where also the relation of the experimental results to
the demands of the quantum theory are discussed in detail (*Phys. Zs.* 20, 132, 1919;
22, 388, 409, 441, 466, 1921). See also the book which has recently appeared by
P. D. Foote and F. L. Mohler (*The Origin of Spectra*, New York, 1922), which contains an excellent survey of the whole region.

of stability of the postulates of the quantum theory, which are foreign to the classical theory. It may be of essential importance for the possibility of setting up such laws of *quantum kinetics* that the laws of the quantum theory for the fixation of the stationary states, as well as for the passage of such states into one another, satisfy the demands for the relativity of the reference system and the reversibility of processes, both characteristic of the classical laws*.

There can in general be no question of a strict application of the classical mechanical laws to the description of the reaction of an atomic system to external effects, except in the special case in which a process of the kind discussed proceeds so slowly and uniformly that the fields of force, which arise from external causes or from other systems concerned in the process, to which the particles of the individual systems are subjected, only change slightly in an interval of the order of magnitude of the periods characteristic of the motions of the particles. In this limiting case, the changes to which an atomic system is subjected during the process, after a common analogy from kinetic theory, may be referred to as an "adiabatic transformation" of the system. The assumption that, in such a case, the reaction of the atom can be described by the use of the usual electrodynamic laws, with the same approximation as in the case of an isolated atomic system, constitutes *the Adiabatic Principle* of Ehrenfest. It must, however, equally be noticed that the application of the principle is naturally limited by the demand, that the motion of the system, if it is to be described by use of classical laws, shall exhibit at each moment during the transformation the periodic properties which are necessary for the fixation of the stationary states, and that the degree of periodicity remain unaltered during the transformation.

* C. Klein and S. Rosseland (*Zs. für Phys.* **4**, 46, 1921) have first shown that the reversibility of non-radiating processes of transition between stationary states is a necessary condition for thermodynamic equilibrium. In this connection they have called attention to the existence of so-called "collisions of the second kind," which play an important part in many phenomena (cf. Franck, *Zs. für Phys.* **9**, 259, 1921). The questions mentioned in the text have recently been discussed by W. Pauli in connection with his detailed investigation of the model for the positive molecular ion H_2^+ (*Ann. d. Phys.* **68**, 177, 1922). He emphasizes the formal applicability of the classical laws in the limiting region of large quantum numbers and concludes from this that the application of the classical laws, even for smaller quantum numbers, in a formal respect gives a certain approximation. From this conclusion, several interesting inferences are drawn. If, however, they are designated as a mechanical correspondence principle, this is a mode of expression which deviates essentially from the conceptions presented in the text. As is explained in Chapter II, the law for the appearance of radiative processes referred to as the Correspondence Principle is to be regarded as a typical law of the quantum theory, which is directly connected with the postulates of the quantum theory and, in itself, is not concerned in principle with the question of the degree of approximation of the applicability of the laws of radiation of the classical theory. An analogy to the Correspondence Principle for non-radiative processes might be introduced by a further development of quantum kinetics, for the laws of which however, as mentioned above, we as yet possess no suitable formulation (cf. Chap. III, § 4).

On the whole the Adiabatic Principle may be regarded as a natural extension of the application of classical electrodynamic laws to isolated atomic systems*. The significance of the Adiabatic Principle in the quantum theory is extraordinarily great, since it leads to the elucidation and development of formal methods for fixing the stationary states. The principle, in fact, demands that the conditions for the stationary states must be of such a kind that they define certain properties of the motion of the system, which will not change during an adiabatic transformation, if the motion is described, with the degree of approximation mentioned, by help of the usual electrodynamic laws. This demand for a so-called "adiabatic invariance" is now in fact fulfilled by the conditions (A), if the quantities *J* on the left-hand side are defined in the manner stated above†. In particular, to conform with this adiabatic invariance, the conditions (A) for a general multiply periodic system may be deduced naturally in a formal way merely by considering the simple special case of a system with independent degrees of freedom, the motion of which for each of these degrees of freedom consists of a simple harmonic vibration, and the stationary states of which are fixed simply by Planck's original formula.

The Adiabatic Principle is further directly applicable to the treatment by the quantum theory of atomic systems which are exposed to the influence of external forces. Thus, the results mentioned above with reference to the changes of the energy of an atom as a result of the presence of a small external field of force can be directly deduced by use of the Adiabatic Principle, as long as the degree of periodicity is not changed by the presence of the forces‡. If, on the other hand the degree of periodicity

---

* See *Q. o. L.*, where an account of the literature on this subject is to be found. The principle of Ehrenfest was referred to there as the principle of the mechanical transformability of the stationary states, to emphasize an essential feature of its content and to avoid a possible confusion with thermodynamic problems (cf. *Q. o. L.* Part I, footnote p. 9).

† For a system, where each motion is simply periodic and the stationary states are thus fixed by the condition that the action integral taken over a period is equal to a whole multiple of Planck's constant (cf. footnote pp. 5, 6), the adiabatic invariance follows, as Ehrenfest has emphasized, immediately from a mechanical theorem of Boltzmann's (cf. *Q. o. L.* Part I, pp. 11–14). For the general case of a multiply periodic system, the adiabatic invariance of the conditions (A) was proved by Burgers in connection with the considerations of Ehrenfest. The method rests on considering the energy function, which appears in the equations of motion (1), to be dependent on one or more parameters besides the independent variables to be used for describing the motion. The transformation is described by the slow change of these parameters. In connection with his proof, Burgers emphasizes the necessity of condition III, which determines the absolute values of the quantities *J*, and proves the insufficiency of the attempts made by several authors to determine these absolute values in principle by an investigation of the limits of the classical mechanical phase-space of the system.

‡ For simply periodic systems it can be shown by a simple elementary calculation that the mean value with respect to the time of the internal energy of the system

increases, from the nature of the case the Adiabatic Principle fails, since the additional conditions (17) for fixing the stationary states naturally cannot be defined on the basis of considerations which rest on classical electrodynamics. We can, however, regard the procedure described in § 3 as a method of selecting the stationary states of a perturbed system from among the possible motions which one would arrive at during the adiabatic increase of an external field of force, if the effect of the external forces, in spite of the change in the degree of periodicity, were calculated simply by the use of classical electrodynamics. We shall return to this point in Chapter II. In summarising, we may say that the Adiabatic Principle ensures the stability of the stationary states in the region in which we might on the whole expect that this stability can be discussed on the basis of the ordinary electrodynamic laws.

The Adiabatic Principle, moreover, helps to overcome a fundamental difficulty of the quantum theory, which concerns the definition of the energy of a stationary state. In § 1, where the energy was formally introduced, there was no mention of continuous changes of energy, and we had as yet no ground on which to anticipate that the differences of energy of the various stationary states, which are important for physical applications, could be calculated, to the necessary degree of approximation, simply by the energy function of the classical theory. This anticipation is so much the less justified, in that a direct process of transference of the system from one stationary state to another cannot even approximately be described by use of the classical laws, a fact implied by our fundamental postulate of the stability of the stationary states. By means of a suitable adiabatic transformation, however, in general at any rate, it is possible in a formal way to achieve by an indirect path a mechanically describable transference of a given stationary state

(that is, the energy in the osculating orbit) does not change during an adiabatic increase of the disturbing forces, to a first approximation, if the motion remains simply periodic (cf. *Q. o. L.* I, 2). In the general case of multiply periodic motions the corresponding theorem is proved in an entirely analogous manner. In fact one readily understands that, in adiabatic increases of the disturbing field of force, the mean value with respect to the time of each quantity remains unaltered, in the definition of which the strength of the external field does not enter, and which, as for example the quantities J, remains constant for the undisturbed motion, while it is subjected, in the disturbed motion, only to small oscillations of a simply or multiply periodic nature.

The last circumstance leads to a simple proof of the adiabatic invariance of the general conditions (A), if one considers the form of the infinitesimal contact transformation which leads from the uniformising variables of the original, to those of the perturbed system. This has come out during a discussion with Dr Kramers. From (11) and (12) it immediately follows that the mean value of the oscillating quantity J_r is equal to the value of the corresponding constant quantity J_r' of a perturbed system, which thus is also equal to the constant value of J_r in the original motion, before the adiabatic increase of the field of force. Since now any adiabatic transformation of the system is resolvable into a great number of infinitesimal contact transformations, it is immediately clear that the adiabatic invariance of the conditions (A) follows.

of a multiply periodic system into another, without, during the process emerging from the domain of the stationary states. In this way, therefore, we are able to define the difference in energy of the two stationary states by means of the classical theory *.

For the purpose of fixing the stationary states, we have up to this point only considered simply or multiply periodic systems. However, as already mentioned in § 1, the general solution of the equations (1) frequently yields motions of a far more complicated character. In such a case, the considerations previously discussed are not consistent with the existence and stability of stationary states, whose energy is fixed with the same exactness as in multiply periodic systems. But now, in order to give an account of the observed properties of the elements, we are forced to assume that the atoms, in the absence of external forces at any rate, always possess "sharp" stationary states, although the general solution of the equations of motion for atoms with several electrons, even in the absence of external forces, exhibits no simple periodic properties of the kind mentioned †. We may obtain a starting point for the discussion of the stationary states of such atoms by a comparison of the motion of the electrons with the interaction of several atomic systems, in particular with a collision of a free electron with an atom. As already emphasized, we must assume that this interaction cannot even approximately be described by use of the classical electrodynamics, and, therefore, it is reasonable to expect a similar result for the motion of the electrons in the atom under their mutual forces, even in the stationary states. That the motion in the stationary states of atoms with several electrons can be described by use of the equations of motion (1), we can rather expect only in particular cases. In the first place, this holds if the interaction of different electrons is of such a kind that it can be compared with an adiabatic influence of several atomic systems, as a result of the great difference in the periods of the various electrons. We encounter a second case, in which a formal application, at any rate, of the conditions of state for multiply periodic systems is possible, in motions of a singular kind, where the various electrons interact in such a way that the motion of each of the electrons taking part in this interaction, as a result of coincidences of periods, is of the type given by the equation (2). In the inter-

---

* See *Q. o. L.* pp. 9 and 20.

† Note added to the proof: In a work which has just appeared by A. Smekal (*Zs. für Phys.* 11, 294, 1922), the view is presented that the motion in the stationary states is always described by such particular solutions of the mechanical equations (1) as can strictly be resolved into harmonic components according to the formula (2). Apart from the difficulties which seem to stand in the way of carrying out this demand in general for atoms with several electrons, it can scarcely be regarded as natural in view of the failure of mechanics in the interaction of atomic systems, emphasized in the text.

vening cases, in which a general coincidence of periods is excluded, without any possible question of adiabatic influence, we must be prepared to admit that the motion of the particles in the stationary states cannot be described by the use of the classical dynamical laws with greater exactness than the exactness with which the motion exhibits simple periodic properties according to these laws. This general failure of the classical laws shows that, even for the case of a harmonic interplay, we must expect that, neither the fixation of the energy, nor the testing of the stability can be strictly carried out by the use of the principles of ordinary mechanics in cases in which the interaction of the electrons cannot be produced adiabatically, or where the effect of the external forces, calculated classically, would change the character of the interaction.

In the following essays, in the discussion of the structure of the atoms of individual elements, we shall go more deeply into this question. We shall try to show that, notwithstanding the uncertainty which the preceding considerations contain, it yet seems possible, even for atoms with several electrons, to characterise their motion in a rational manner by the introduction of quantum numbers. In the fixation of these quantum numbers, considerations which rest on the Adiabatic Principle, as well as on the Correspondence Principle discussed in the next chapter, play an important role. The demand for the presence of sharp, stable, stationary states can be referred to, in the language of the quantum theory, as a general principle of *the existence and permanence of the quantum numbers.*

§ 5. *The Statistical Weights of the Stationary States.* Before we close the general consideration of the stationary states we must say a few more words on the statistical applications of the quantum theory. The central problem here is to determine the "weight" which is to be attributed to the various states in the calculation of the probability of a statistical distribution of a number of atoms over all possible stationary states, which according to Boltzmann's principle controls the investigation of thermodynamic questions. Ehrenfest has made a contribution to this question of decisive importance. By an investigation of the conditions of validity for the statistical basis of the second law of thermodynamics, he has deduced a condition, which, when applied to our main postulate of the existence of separate stationary states in an isolated system, immediately asserts that the weight which is to be attributed to any individual stationary state, defined by the quantum numbers $n_1, ..., n_u$, is the same for two systems if the sets of stationary states of these systems can be connected, without ambiguity, by a continuous transformation*. This law can be used to determine

* See *Q. o. L.* Part I, p. 11. Since the original treatment by Ehrenfest (*Phys. Zs.* 15, 660, 1914) is not adapted to the form of the quantum theory discussed here,

the statistical weights of the stationary states of a given atomic system, if the weights of the states of a system are known, which can be transformed continuously into the given one. For multiply periodic systems one is led to the conclusion, by considerations of this kind, that the weights of the various stationary states of a non-degenerate system must have the same value, which can be put equal to h^r, where r represents the number of degrees of freedom of the system*. This determination is not only supported by ex-

which rests on the postulate of the existence of discrete stationary states, but deals with the possibility of a continuous distribution of the states of motion in the phase-space, it may be useful to give a deduction of the condition for the invariance of the statistical weights, mentioned above, which takes a very short and clear form, if it is based directly on our fundamental postulate. We consider a great number N of like atoms. We will designate their various stationary states by a suffix τ, and the energy and the statistical weight of a definite state by E_τ and g_τ respectively. The probability of a statistical distribution, for which the number of atoms in the τth state is N_τ, is then, of course, given by the expression:

$$W = N!\, \Pi_\tau \frac{g_\tau^{N_\tau}}{N_\tau!}.$$

That distribution for which W is a maximum for a given value of the total energy

$$E = \Sigma_\tau N_\tau E_\tau$$

is further determined by

$$N_\tau = C g_\tau e^{-\frac{E_\tau}{kT}},$$

where k is Boltzmann's constant and T the absolute temperature, and in which C is further determined by the condition

$$\Sigma_\tau N_\tau = N.$$

For the entropy S of the system we find, according to Boltzmann's relation

$$S = k \log W,$$

by use of Stirling's formula, the following expression:

$$S = k\Sigma_\tau N_\tau \log g_\tau - k\Sigma_\tau N_\tau \log N_\tau + kN \log N.$$

We now consider a thermodynamic process in which each atom will be subjected to the same transformation, in the sense, that they are all exposed to the same external forces, and we assume that work δA is taken from the system as a whole and a quantity of heat δQ is added to it. Thus we have:

$$\delta Q = \delta E + \delta A = \delta \Sigma_\tau N_\tau E_\tau - \Sigma_\tau N_\tau \delta E_\tau = \Sigma_\tau E_\tau \delta N_\tau.$$

Here δE_τ is the change of the energy of the corresponding stationary states as a result of the transformation. It is to be noted that the validity of the equation for δQ is not connected with the assumption that the behaviour of the atoms during the transformation is describable by the laws of classical mechanics, but that this equation is to be regarded as an immediate result of the application of the conception of energy to the transformation.

According to the second law of thermodynamics we have now, on the other hand,

$$\delta Q = T\delta S = kT\Sigma_\tau \log \frac{g_\tau}{N_\tau}\, \delta N_\tau + kT\Sigma_\tau \frac{N_\tau}{g_\tau}\, \delta g_\tau,$$

which, by use of the above equations, can also be written:

$$\delta Q = \Sigma_\tau E_\tau\, \delta N_\tau + CkT\Sigma_\tau e^{-E_\tau/kT}\delta g_\tau.$$

By a comparison of this expression for δQ with the above, it follows that the last member must vanish for all temperatures, which is only possible when $\delta g_\tau = 0$, in agreement with the assertion in the text.

* The determination of the statistical weights of a degenerate system is gone into in more detail in Q. o. L. pp. 35–37, 107, 133. Here it can only be recalled that on the basis of thermodynamic stability the statistical weights of the stationary states of a degenerate system can be determined by considering a family of non-

perimental facts on the specific heats at low temperatures, the explanation of which, as is well known, rests on the application of the quantum theory to simple mechanical systems with several degrees of freedom, but it also indicates that the statistical application of the quantum theory to temperature equilibrium follows asymptotically the application of classical mechanics in the region of large quantum numbers, where the motions in the neighbouring states are relatively only slightly different. According to the classical theory, the *a priori* probability that the representative point of a mechanical system lies within a certain region of the phase-space, is equal to the volume of this region. If we now inquire how large the region is which corresponds to the motions which belong to the values of the quantities J, which lie between J_r' and J_r'', we then find, according to the definition of these quantities, that this region is equal to $\Pi (J_r' - J_r'')$*. If we now, on the other hand, consider the region of large quantum numbers and put $J_r' = n_r'h$ and $J_r'' = n_r''h$, then $\Pi_r (n_r' - n_r'')$ quantum states are contained in the region considered, and with each stationary state, therefore, there can be correlated a region of the phase-space of magnitude h^r.

Up to this point we have considered the question of the weights of the stationary states primarily in their relation to the statistical applications of the quantum theory. Naturally, however, these weights express, on the other side, properties of the stationary states, which are not necessarily connected with the question of statistical distributions. We meet this particularly when we turn to the problem of atomic structure. Here one is led to exclude certain conceivable quantum states, and thus to ascribe to them the weight zero. This concerns not only the cases where the more exact consideration of the corresponding motion already indicates that it is to be regarded as unsuited for a stationary state, but, by the law of the invariance of the weights in continuous transformations, based on thermodynamics, we are led also to exclude all those conceivable quantum states which can be brought over by a

degenerate systems, which contains the degenerate system as a limiting case. To each stationary state of the degenerate system a weight must be attributed which is equal to the sum of the weights of those states of the non-degenerate systems which pass over into it in the limiting case. The demand that this sum should have the same value for all families of non-degenerate systems, which contain the given degenerate system as a limiting case, offers in certain cases a support for the exclusion of individual conceivable quantum states. This will be mentioned in the text farther on.

* This follows simply from the circumstance that the quantities J_r, w_r arise from the quantities p_r, q_r by a contact transformation, and, that, according to a well-known law of mechanics, in such a transformation the volume-element of the phase-space retains its extension. Thus we have:

$$\Pi_r dp_r dq_r = \Pi_r dJ_r dw_r,$$

from which the above expression follows directly by considering the definition of the uniformising variables w_r. (See in this connection, J. M. Burgers, *Dissertation*, p. 254.)

continuous transformation into one of the singular states just mentioned*. We shall go into this question more closely in special cases in the following essays.

The considerations in these paragraphs may further point out the direction in which a formulation of the laws of quantum kinetics is to be sought, of which mention was made at the beginning of the previous section. The existence and the stability of the stationary states can, in fact, be formally so conceived that, of the kinematically conceivable possibilities of motion, only these states will have a weight-function differing from zero. Now we may demand that the laws which govern the frequency (probability, cf. Chap. III) of the process of transference in the interaction of atomic systems must pass over into the continuous laws of classical mechanics in the region of large quantum numbers, when we introduce the expressions for the corresponding phase-space, in place of the discontinuous weight-function just mentioned. This demand may perhaps provide a point of view, in analogy with the consideration in Chapter II, § 2, for seeking out these laws of quantum kinetics.

CHAPTER II
THE PROCESS OF RADIATION

§ 1. *The Second Fundamental Postulate.* The second postulate of the quantum theory for isolated atomic systems characterises more exactly the relations for the exchange of energy between an atom and a field of electromagnetic radiation. According to the first postulate, such an exchange only takes place during processes which can be described as complete transitions between two stationary states. The second postulate now asserts that every emission of radiation connected with such a process consists in the emission of *purely harmonic waves*, the frequency ν of which is given by the so-called *frequency-condition*,

$$h\nu = E' - E'', \qquad \ldots\ldots(B)$$

where E' and E'' represent the energy of the atom in the two stationary states. It is further required that every process of absorption, in which the atom is carried from one stationary state to another by the effect of electromagnetic radiation, is conditioned by illumination with waves, whose frequency is given by the same relation (B).

The content of this postulate is, in many respects, such as to sharpen the break with classical electrodynamics, which the first

---

* See *Q. o. L.* Part I, p. 37; Part II, p. 107.

postulate introduced. According to the classical theory, as has been mentioned, every motion of the particles of an atomic system gives rise to the emission of electromagnetic radiation. The nature of this radiation, at any rate to a first approximation, is determined by the variation of the total electric moment of the system with the time. In those isolated systems in which the motions, apart from the reactions due to radiation, possess such periodic properties as seem necessary for the fixation of the stationary states, the displacement of any particle can be regarded, according to the expression (2), as compounded of a number of harmonic vibrations. Hence the radiation at any moment, to a first approximation, may be regarded as composed of a number of wave-systems, the frequencies ν of which are each equal to one of the frequencies $\tau_1\omega_1 + \ldots + \tau_u\omega_u$ appearing in the motion. The intensities are determined by an expression of the form:

$$\Delta E = \frac{2}{3}(2\pi)^4 \frac{e^2}{c^3} \nu^4 \overline{A^2} \Delta t, \qquad \ldots\ldots(19)$$

where ΔE is the energy emitted in an element of time Δt, and $\overline{A^2}$ is the mean value of the square of the vibrational displacement of the harmonic component of the electric moment belonging to the corresponding frequency. Further, according to the classical theory, every exchange of energy between an atom and a field of radiation will be conditioned by the presence of wave-systems in this field, the frequencies of which very nearly coincide with the frequencies of certain of the harmonic components of the electric moment. The result of this exchange will not only depend on the amplitudes of these wave-systems and of the corresponding components of vibration, but also on the phase-difference between them, and of course, in such a way that the atom will receive or give up energy according to the value of this phase-difference.

According to the postulates of the quantum theory, we not only abandon any such immediate connection between the motion of the atom and the result of the process of emission or absorption, but we are even compelled to depart so far from the ordinary descriptions of nature as to assume that the result of such a process depends actually on the final, as well as the initial state. This relation perhaps stands out most clearly at present for the process of emission, since, on the basis of the postulates, one and the same stationary state of the atom can serve as the origin of quite different radiative processes. These correspond to transitions from this state to various other stationary states. Furthermore, in the presen state of the theory, it is not possible to bring the occurrence o radiative processes, nor the choice between various possible transitions, into direct relation with any action which finds a place in our description of phenomena, as developed up to the present time. Under these circumstances, we are naturally lead to the method of treatment which was first applied by Einstein in his deduction of the laws of temperature radiation on the basis of the postulates of the quantum theory in the form given here. According to this method of treatment, we do not seek a *cause* for the occurrence of radiative processes, but we simply assume that they are governed by *the laws of probability*. Thus we shall assume, with Einstein, that an atom in a stationary state possesses a certain probability of shifting to another state of smaller energy within a given element of time, with the emission of radiation. This occurs spontaneously; that is, without any assignable external stimulation. Exactly as in radio-active processes, this probability is taken as proportional to the element of time Δt. The proportionality factor, the so-called coefficient of probability, is characteristic of the process of transition in question, and depends only on the nature of the system. In this connection, Einstein emphasizes the formal analogy which this assumption exhibits with the conceptions of ordinary electrodynamics, in spite of its estranging character. According to these conceptions, of course, there is no mention of the laws of probability, but, as described above, the radiation even here is conditioned solely by the system itself, and not by external causes. The further assumptions introduced by Einstein concerning the effect of an external radiation on an atomic system are closely related to this analogy. According to the classical theory, as mentioned, illumination (Bestrahlung) of an atom with waves of a frequency which nearly coincides with a characteristic frequency in the motion of the system, will lead to an increase or a decrease of the energy of the system, according to the phase-difference between the waves and the motion of the atom. In analogy to this, Einstein assumes that the result of illuminating an atom with waves, the frequency of which fulfils the relation (B), can be described by saying that atoms which possess an energy E'' will acquire a probability for a transition to a state of greater energy E' by the absorption of a quantity of energy equal to $h\nu$. At the same time, atoms with energy E' acquire an additional probability for a transition to the state with energy E'' by the emission of radiation with a frequency ν. In both cases the coefficient of probability is assumed to be proportional to the energy density of the radiation for the frequency in question.

In spite of the great importance which undoubtedly belongs to Einstein's method of treatment in view of its results, in many respects it can be regarded only as a preliminary solution. This appears already in the formulation of the assumptions, in which the duration of the process of transition is not taken directly into account, although its magnitude, as we shall see later on, plays an essential rôle in the description of phenomena. Further, that the assumptions are approximate is also a consequence of the fact that, in atomic systems which are exposed to radiation so intense that the external forces are no longer small compared with the forces acting between the particles in the undisturbed system, a description of the stationary states cannot be accomplished without reference to the forces of the field of radiation. This last point leads us again to the treatment of non-isolated systems given by the quantum theory. We shall come back to this later in this chapter.

§ 2. *The Correspondence Principle.* Notwithstanding the fundamental break with the classical electrodynamic theory introduced by the postulates of the quantum theory, it has yet been possible, on the basis of the quantitative conditions (A) and (B), to connect the occurrence of radiative processes with the motion in the atom in a way which offers an explanation for the fact that the laws of the classical theory are suitable for the description of the phenomena in a limiting region. This is attained, if the various possible radiative processes are correlated with the harmonic vibrational components, appearing in the motion of the atom, This is to be done in such a way that the possibility of the occurrence of a transition, accompanied by radiation, between two states of a multiply periodic system, of quantum numbers for example $n_1', ..., n_u'$ and $n_1'', ..., n_u''$, is conditioned by the presence of certain harmonic components in the expression given by (2) for the electric moment of the atom. The frequencies $\tau_1 \omega_1 + ... + \tau_u \omega_u$ of these harmonic components are given by the following equations:

$$\tau_1 = n_1' - n_1'', ..., \tau_u = n_u' - n_u''. \qquad ......(20)$$

We, therefore, call these the "corresponding" harmonic components in the motion, and the substance of the above statement we designate as the "Correspondence Principle" for multiply periodic systems*.

It is well known that one obtains a clue to the law mentioned for the occurrence of radiative processes just by investigating the conditions necessary in order that the description of the phenomena on the basis of the postulates may be asymptotically in agreement, in the limiting region of large quantum numbers, with the results of the classical theory, the applicability of which in this region to statistical problems seems secure. If we consider a multiply periodic

---

* In *Q. o. L.* this designation has not yet been used, but the substance of the principle is referred to there as a formal analogy between the quantum theory and the classical theory. Such an expression might cause misunderstanding, since, in fact—as we shall see later on—the Correspondence Principle must be regarded purely as a law of the quantum theory, which can in no way diminish the contrast between the postulates and electrodynamic theory.

system, the stationary states of which are fixed by the conditions (A), we obtain, on the basis of the relation (B) and with the help of (7), for the frequency of the radiation which is emitted on a transition from one stationary state with quantum numbers $n_r{}'$ to another with quantum numbers $n_r{}''$, $(r = 1, ..., u)$, the expression

$$\nu = \frac{1}{h}\{E' - E''\} = \frac{1}{h}\int \Sigma \dot{\omega}_r \, dJ_r. \qquad \ldots\ldots(21)$$

If now the quantum numbers $n_r{}'$ and $n_r{}''$ are large compared with their differences $n_r{}' - n_r{}''$, and, as a result, the motions in these two states differ comparatively little from each other, the frequencies under the integral sign can be regarded as approximately constant during the integration. Thus we obtain by use of equations (A) the asymptotic relation:

$$\nu \sim \sum_1^u (n_r{}' - n_r{}'')\,\omega_r.$$

One thus sees that the frequency ν of the radiation coincides asymptotically with the frequency $\tau_1\omega_1 + \ldots + \tau_u\omega_u$ of a harmonic component appearing in the motion of the system according to (2), and, of course, with that one for which the relations (20) are fulfilled.

As regards the frequency of occurrence of the various possible transitions, we are at once led to the assumption that the intensity, with which the various spectral lines in this limiting region appear, is determined by the amplitude of the corresponding component of vibration in the electric moment of the system, approximately in the same way as would be the case in classical electrodynamics according to formula (19). We arrive at this conclusion from the stated connection, in the region of large quantum numbers, between the frequencies of the components of the motion and those of the wave-trains which are emitted on the transition. One is, therefore, led to the conception that the occurrence of radiative transitions is conditioned by the presence of the corresponding vibrations in the motion of the atom. As to our right to regard the asymptotic relation obtained as the intimation of a general law of the quantum theory for the occurrence of radiation, as it is assumed to be in the Correspondence Principle mentioned above, let it be once more recalled that in the limiting region of large quantum numbers there is in no wise a question of a gradual diminution of the difference between the description by the quantum theory of the phenomena of radiation and the ideas of classical electrodynamics, but only of an asymptotic agreement of the statistical results. As we shall see, the applicability of the principle for elucidating problems of the quantum theory is primarily connected with just this point.

If we inquire into the absolute values of the coefficients of probability, appearing in Einstein's theory of temperature radiation, which quantitatively determine the occurrence of radiative transitions, it must be emphasized, on the other hand, that the above considerations naturally only permit us to calculate these coefficients simply by means of the amplitudes of the corresponding harmonic components of the motion in the region of large quantum numbers. This is quite clear from the fact that only in this limiting region are the amplitudes in the initial and the final states approximately equal. Just as is the case for the frequencies of these components, we must in general be prepared for the fact that the corresponding amplitudes in the two states may be entirely different. The possibility, however, does not seem to be excluded of finding a general expression for the said coefficients of probability by the use of mechanical symbols. An indication in such a direction is perhaps given by remarking that the frequency associated with a transition between any two stationary states of a multiply periodic system can be expressed as a simple mean value of the frequency of the corresponding vibrations in a continuous series of states of motion, suitably chosen from the general solution of (1). We will again consider a transition between two stationary states, for which the quantum numbers in the equations (A) are respectively n_r' and n_r''. Then we shall consider the states, in which the quantities representing the uniformised momenta $J_1, ..., J_u$, which serve to fix the stationary states, are given by

$$J_r(\lambda) = h \{n_r'' + \lambda (n_r' - n_r'')\},$$

where the parameter λ can assume all values between 0 and 1. As is immediately seen from (21) and (A), one can then write the expression for the frequency ν of the radiation emitted on a transition in the following manner:

$$\nu = \int_0^1 \Sigma (n_r' - n_r'') \omega_r (\lambda) d\lambda.$$

The frequency of the wave-system emitted on a transition can, therefore, be regarded as a mean value of the frequencies of the corresponding vibration in the series of "intermediate states" under discussion. Kramers has called attention to this simple relation in a communication which contains a detailed investigation of the application of the Correspondence Principle to the question of the intensity of spectral lines. He also discusses there the possibility of finding a general expression for the probability of a transition by use of a suitable mean value over the intermediate states of the quantities which, according to the classical theory, determine the radiation of energy accompanying the corresponding vibration in the electric moment of the atom*.

---

* H. A. Kramers, *Kgl. Danske Vidensk. Selsk. Shrifter*, 8 Raekke, III. Although this important question is not yet solved, still it must be remarked that the argument deduced from the "spectroscopic stability" disproving the validity of the expression set up on p. 100 of Kramers' essay, cannot be maintained. (See p. 27 of this essay.)

§ 3. *The Correspondence Principle and the Fixing of the Stationary States.* If the Correspondence Principle cannot instruct us in a direct manner concerning the nature of the process of radiation and the cause of the stability of the stationary states, it does elucidate the application of the quantum theory in such a way that one can anticipate an inner consistency for this theory of a kind similar to the formal consistency of the classical theory. Firstly, the previously mentioned rôle of the periodic properties of the motion for fixing the stationary states comes clearly to light. Then the assertion that the number of quantum conditions (A) is exactly equal to the degree of periodicity, becomes a necessary demand for attaining an unambiguous correspondence between the various types of transitions and the harmonic components appearing in the motion. The addition of further conditions, if the degree of periodicity increases under the influence of external forces, appears too in a very simple light. We can, in fact, regard these conditions as an immediate demand for a correspondence between the new, slow harmonic vibrations appearing in the secular perturbations and processes of transition, for which the quantum numbers already appearing in the undisturbed motion are not changed, but only the new quantum numbers, appearing in the additional conditions*.

It is also of interest in this connection to mention that light is thrown by the Correspondence Principle on certain apparent paradoxes which we meet in fixing the stationary states of strictly simply or multiply periodic systems, in which the motion, within periods of the same order of magnitude as the fundamental periods of the motion, exhibits approximately such periodic properties, as, considered by themselves, would lead to an entirely different fixation of the stationary states from that to which one would be led by the procedure of Chapter I, § 2, if one takes account of the strict periodic properties of the system. In order to facilitate the discussion, we shall designate the strict periods of the system as the "macroperiods", while we shall call the periods which reveal the quasi-periodic properties, the "microperiods", since the apparent paradoxes only appear if these latter periods are very short compared with the former. On the basis of relations (A) and (B) one might at first think that we might meet with a singular difference from the demands of the classical theory, since, of course, the microperiodic properties of the motion would apparently not be recognizable in the spectrum at all. This, however, is by no means the case. For if one more closely considers the probability of the various quantum jumps, we find that certain jumps, in which the quantum numbers of the states fixed by the macroperiods change

---

* *Q. o. L.* II, 2, p. 58. Examples in which this point of view can be used with special simplicity are presented by the problems of the effect of external electric and magnetic fields on the hydrogen lines. See *Zs. für Phys.* **2**, 423, 1920, also reprinted in *The Theory of Spectra and Atomic Constitution,* C. U. Press, 1922; see also *The Seventh Guthrie Lecture, Proc. Phys. Soc. London,* **35**, 275, 1923.

by many units, are especially probable, as a result of the appearance, occasioned by the microperiods, of certain overtones, or groups of overtones, with particularly large amplitudes. In this way the microperiods appear in the spectrum in a way quite similar to that in which they would appear in the radiation according to the classical theory. This application of the Correspondence Principle, to which Ehrenfest and Breit* have called attention in a work which has recently appeared, on the whole expresses clearly the close connection between radiation and motion in the quantum theory, which persists in spite of the fundamental difference between the character of the postulates and the continuous description of the classical theory. A peculiar difficulty, which is encountered in the limit, in which the macroperiods are extraordinarily long compared with the microperiods, and to which the authors mentioned have called attention, appears to have its natural foundation in the limitation of the validity of the postulates and will be discussed in more detail in the next paragraph (see p. 30).

The light which is thrown by the Correspondence Principle on the fixing of the stationary states appears to give us further guidance, as already indicated in § 3, on the question of fixing these states for systems, in which, as for atoms with several electrons, the general solution of the equations (1) possesses no simple periodic properties. That is, the principle offers a basis for reducing the possibilities of processes of transition connected with the emission of radiation, during the building up of an atom by the successive binding of electrons, as well as during the reorganisation of an atom after a change of electronic configuration occasioned by external influences. In the following essays we shall return to this question in more detail. We shall particularly try to show that the principle offers a point of attack on the problem of the stability of the normal states of the atom, fundamental to the discussion of the properties of the elements.

§ 4. *The Correspondence Principle and the Nature of the Radiation.* In setting up the Correspondence Principle it is essential to assume that an intimate connection exists between the observable character of the radiation emitted on a transition between the stationary states according to the second postulate, and the radia-

---

* P. Ehrenfest and G. Breit, *Zs. für Phys.* **9**, 207, 1922. As an example, the authors used a system which consists of a particle able to move freely in a circular orbit. However, it is subjected to the further condition that, after a certain number of revolutions in any direction, the direction of rotation reverses. The free revolution is here the microperiodic motion, while the regular reversal represents the macroperiodic properties. The authors show how the free rotation gives rise to the appearance of certain overtones with large periods in the periodic motion determined by the regular reversal of the direction of rotation. These overtones cause a preference for transitions which correspond to great changes of the stationary states of the last-named motion. With regard to the change of energy of the system, these transitions correspond approximately to transitions between the stationary states of a freely rotating particle.

tion which would be emitted according to classical electrodynamics on account of the presence of the corresponding vibrational component in the electric moment of the atomic system. In general we shall thus expect that the nature of the radiation, observed in various directions, will be the same as that which would be emitted by an electron, according to the classical theory, which describes an elliptical harmonic vibration. In the cases where the corresponding vibration for each motion of the system is linear or circular, as is the case in a non-degenerate system with axial symmetry, we shall thus expect that the system of waves emitted will exhibit plane or circular polarisation. These consequences have been confirmed in all cases where it has been possible, as for the action of electric and magnetic fields on spectral lines, to compare them with experiment. In this connection it might be of interest to call attention to the fact that, in spite of the close connection between radiation and motion even in the question of polarisation, still we must be prepared in certain respects to encounter pronounced deviations from the classical theory. Much as it is the postulates of the quantum theory which allow us at all to expect sharp spectral lines for atomic systems, the peculiar relations of stability of the stationary states and the character of the radiation emitted during a process of transition, compel us to expect in certain cases a discontinuous change of the polarisation where the classical theory does not require it. We obtain a characteristic example of this if we bring an isolated atomic system into a magnetic or electric field. While, according to the classical theory, each orientation of the atom as a whole with reference to the field will, to a first approximation, be equally frequent, a different state of affairs exists according to the quantum theory. As we mentioned, the additional periods of the motion, arising from the secular perturbations, will require special conditions for the stationary states, so that certain orientations are given preference*. Besides the characteristic polarisation

* A direct, non-spectroscopic proof of the existence of such conditions of orientation has, of course, been obtained by O. Stern and W. Gerlach (*Zs. für Phys.* 9, 349, 1922) in their beautiful and important investigation on the deflection of moving atoms of silver in a non-homogeneous magnetic field. The question as to the rapidity with which the orientation of the atoms is attained in these experiments has been discussed in detail in a note which appeared recently, by A. Einstein and P. Ehrenfest (*Zs. für Phys.* 11, 31, 1922). In connection with the fundamental difficulties standing in the way of a detailed description of this adjustment, which are discussed in the above note, it may be pointed out that the effect of the field on the harmonic components, into which the motion of the atom may be resolved, consists not only in the addition of new vibrations, the frequencies of which are proportional to the external forces, but also in the introduction of a change in the harmonic components of the motion, already present in the undisturbed atom. The velocity of readjustment of the atom in the field can scarcely, therefore, be estimated from the life of the stationary states of an imaginary atom, in the motion of which only the first mentioned vibrations were present, as is attempted in the note under discussion. On the contrary, the life of the stationary states of the excited, undisturbed atom might be regarded as determining this velocity of

28   *The fundamental postulates of the quantum theory*

of the various components into which the individual lines are resolved, we must be prepared, in contrast to the classical theory, for the fact that the total light of the components, even at very weak fields, can exhibit a characteristic state of polarisation relative to the axis of the field. The circumstance that such an effect seems to be confirmed by various observers* might, therefore, be regarded as a support for the assumptions of the quantum theory, similar in kind to the fact that in general sharp spectral lines are observed†.

We wish to mention in these paragraphs a few more questions which are immediately connected with the formulation of the second postulate, on which we have intentionally not yet touched. I am thinking here partly of the question as to the frame of reference in which the frequency ν of the wave-system, emitted by a process of radiation, must be measured, partly of the problem of the sharpness of definition of this frequency. As regards the first point, the Correspondence Principle, in the case of isolated systems, leads immediately to the assumption that the frame of reference for the measurement of the frequency in the relation (B), much as was the case for fixing the stationary states by the conditions (A), must be so chosen that the system as a whole is at rest in it‡. Naturally we shall assume that in another frame of reference the wave-system emitted, observed in various directions, exhibits a Doppler-effect of the kind well known from the theory of relativity, as has also been found in the well-known experiments with canal rays. For non-isolated systems, however, we meet with difficulties, since a fixed, definite, natural system of reference does not present itself.

readjustment. That, in the experiments mentioned, we have to deal with the undisturbed motion, not of excited atoms, but only of atoms in the normal state, does not in principle stand in the way of such a conception. On the contrary, it is only the formal nature of the quantum theory in its present form (see Chapter III) that stands out here with particular clearness.

* See W. Voigts' article on magneto-optics, Grätz, *Handbuch der Elektrizität*, IV, p. 624. See also H. Rausch von Traubenberg, *Naturwissenschaften*, **10**, 791, 1922, who has recently determined the corresponding effect of a magnetic field in the particularly simple case of the spectrum of hydrogen.

† The assumption that the polarisation of the total light of a spectral line cannot be essentially changed by the influence of weak external fields was regarded in *Q. o. L.* as a necessary demand of spectroscopic stability. Since this demand cannot be regarded as justified, according to the fundamental ideas of the quantum theory, a fundamental argument against the possibility of representing universally the probability of transitions between stationary states by a simple expression by means of mechanical symbols breaks down.

‡ This requires naturally that this system of reference is the same before and after the process of radiation. An uncertainty on this point may arise if we would assume that the process of radiation is connected with a change of momentum. The question is discussed in an interesting way by Schrödinger (*Phys. Zs.* **23**, 301, 1922) in connection with the idea put forward by Einstein, that the emitted radiation is entirely directed (see Chapter III). Apart from the fact that this idea is far removed from the presentation here given of the actual applications of the quantum theory, it may be recalled that, simply because of the smallness of the mass-ratio of the negative and positive particles of the atom, an eventual change of momentum can have no observable effect on the spectra of isolated systems.

A characteristic example, which will be treated in the next paragraphs, is offered by a consideration of a collision between an atomic system and a free electron, which is accompanied by radiation, without, however, the electron becoming bound by the atom. The second of the above-mentioned problems relates to the sharpness with which the frequency of the emitted waves is defined. Purely kinematically, the finite duration of the process of radiation sets an upper limit for this sharpness. According to the requirements of correspondence it is reasonable, at least for isolated systems, to assume that an upper limit for the time, as far as order of magnitude is concerned, is given by the time in which, according to the classical theory, a corresponding quantity of energy would be emitted by a vibrating electron, the frequency of which is equal to that of the radiation, and the amplitude of which is of the order of magnitude of the corresponding vibration*. One sees that the limit for the sharpness of definition of the frequencies of the waves, to which we are led in this manner, is exactly the same as the approximation with which we can expect to calculate, by the use of (B), the frequency of a radiation emitted by a process of transition, if one considers that the description of the motion in the stationary states, and the determination of the energy by means of conditions (A), already includes the neglect of the radiation-reactions. As several times emphasized in the previous chapter, the approximate character of such a description arises from the fact that the reactions due to radiation were neglected in the application of the equations of motion (1). There is a question here not only of our ignorance of the modifications which are to be intro-

---

* (See *Q. o. L.* II, p. 94, note.) It is well known that this limit of sharpness of spectral lines is in approximate agreement with the upper limit for the time of radiation which can be deduced from the gradual cessation of the light from canal rays (see W. Wien, *Ann. d. Phys.* 60, 597, 1919, and 66, 229, 1921). This problem is briefly discussed by A. Sommerfeld and W. Heisenberg (*Zs. für Phys.* 10, 393, 1922), starting from the same general point of view as is indicated in the text. These authors try to obtain a more exact theoretical estimation of the breadth of spectral lines. Although this attempt contains interesting and promising ideas, it is very difficult in the present state of the theory to decide how far a definite, quantitative method of estimation can be necessarily inferred from the Correspondence Principle. The same can be said of the interesting work of G. Mie (*Ann. d. Phys.* 66, 237, 1921) in connection with the experiments of Wien, according to which the intensity of the waves during the process of radiation should at first gradually increase, and, after attaining a maximum, again gradually decrease at the end of the process, and which can find, in the present state of the theory, apparently no direct foundation from the point of view of the Correspondence Principle. It might still be of interest to note the circumstance which serves Mie as a starting point. It is that in the spectral line H_α particularly investigated the frequency of the radiation emitted is equal to the frequency of revolution of the electron in a conceivable solution of the equations of motion, which can be regarded as lying between the two states which act as initial and final points for the process. This is a special case of a rule which, for multiply periodic systems, has general validity. According to this rule, the frequency of the radiation emitted can be regarded as a mean value of the frequency of the corresponding vibration over a continuous series of conceivable "intermediate states" (see above, p. 24).

duced into the electrodynamic laws in order to account for the absence of the emission of radiation in the stationary states, but also the kinematical problem regarding the periodic properties of the motion is even undetermined, in the sense that we have to reckon with the interruptions of the regular motion which correspond to the processes of radiation, whether they are spontaneous transitions to states of lower energy, or forced transitions occasioned by external radiation.

This lack in sharpness of the description of the motion of the electrons in the atom brings a lack of sharpness into the definition of the stationary states, the consideration of which in certain cases possesses essential significance. If we consider the example treated at the end of the preceding paragraph, we encounter a new situation when the macroperiods are so great that their length becomes of the same order of magnitude as an interval of time in which the probability is not small that one of the quantum jumps occasioned by the properties of the microperiodic motions has taken place. In this case the strict application of the equations of motion (1) to the description of the eventually macroperiodic properties becomes illusory, and we are led to the conception that in the limit every effect of such properties on the fixation of the stationary states vanishes, and also on the observable properties of the system, which now can only depend on the so-called microperiodic properties. By such a consideration one seems in a natural way to overcome the difficulties emphasized by Ehrenfest and Breit (*loc. cit.* p. 210), which relate to an unambiguous determination of the thermal properties of such a system. These difficulties consist in the fact that the quantities important for the temperature equilibrium, that is, the values of the energy and of the statistical weights of the stationary states, will depend, in a formal consideration of the strict solution of the equations of motion (1), exclusively on the macroperiodic · properties of the motion in this solution. One, therefore, apparently arrives at a discontinuity if one goes to the limit, where strictly only microperiodic properties appear. From the above considerations it now follows, however, that one must expect that, even before the limit of infinitely long macroperiods is attained, their effect on the observable properties will have gradually vanished. It should be emphasized in this connection that, for the decision whether in a given case the micro- or the macroperiods are essentially important, the actual temperature must be taken into account, because of the effect of temperature radiation on the frequency of occurrence of the forced quantum jumps.

§ 5. *The Radiation from Non-Isolated Systems.* In the consideration of processes of radiation, we have up to this point only considered those in which isolated atomic systems were concerned, the stationary states of which can be fixed by the conditions (A).

In the consideration of non-isolated systems, where we shall specially consider the interaction of two or more atomic systems, we meet with essentially different relations, although they can be investigated, to a certain extent, on the basis of the postulates. As we already mentioned in the previous chapter, the first postulate demands that two atoms which interact shall exist in stationary states, after, as well as before, the process. In this connection it is important to recall the fact that the stationary states are fixed solely through the relative motions of the particles in each atom, so that the requirement stated expresses nothing concerning the relative motion of the two atomic systems after interaction, if they have widely separated. Only when the interaction leads to a combination of the systems, can we expect that the motion, apart from the motion of the system as a whole, will be fixed by conditions of the type (A). A simple example of such a process is given by the "collision" between a free electron and a positive atomic ion, which under certain circumstances can lead to the binding of the electron by the atom with the emission of radiation. From the point of view of the second postulate, we then have in such a process a possibility of a further application of the Correspondence Principle in the fact that the change of the electric moment of the total system with respect to the time, if the motion is approximately described by equations (1), contains harmonic components of all possible frequencies. This means that the moment can be represented by a Fourier integral, instead of a trigonometric series of the form (2). This corresponds to the circumstance that we can imagine here, on the basis of the indefiniteness of the relative motions of the two atoms before and after the process, a radiative process where the frequency of the radiation, calculated according to (B), can assume all possible values continuously distributed over the interval. We thus obtain a possibility of a formal interpretation of the so-called continuous spectrum in a way which exhibits an analogy to the views of the classical theory, corresponding to the explanation by the quantum theory of the line spectra of isolated atomic systems of a multiply periodic character*. The difference between the assertions of the two theories is quite clearly apparent even here, since, of course, the quantum theory at the same time offers an explanation for the appearance of a sharp limit for the range of frequency on the short wave-length side, which, for example, is observed in the X-ray region, when the anticathode is bombarded with electrons with a given velocity. As is known,

* *Q. o. L.* II, p. 137. In an investigation which will soon be published, Kramers has shown that it is possible to make a quantitative use of such a point of view, since he has succeeded in giving an interpretation by the quantum theory of the empirical laws for the absorption of homogeneous X-rays in their characteristic dependence on the wave-length and the atomic number of the absorbing element. [Translator's Note. See *Phil. Mag.* 46, 836, 1923.]

the limiting frequencies are given directly by the relation (B), if we introduce the kinetic energy of the electrons for $E' - E''$. We scarcely need to recall in this connection that this phenomenon forms the "inverse" of the photoelectric effect, as it is described in Einstein's well-known theory. Such an application of equation (B) to the process of radiation seems natural just in a determination of this limit, in which the relative velocity between the electron and the atom has vanished after the interaction, since we have to do with a phenomenon which can be immediately reversed by the absorption of monochromatic radiation. The question as to the strict validity of the second postulate for an interaction between an electron and an atomic ion, which does not lead to binding, gives rise to difficulties of a different kind. In an attempt to apply the condition (B) one is at once confronted with the question touched on in the previous paragraph concerning the frame of reference in which the frequency of the wave-train is to be measured. With reference to the classical theory it seems plausible, since, of course, the electron, because of its small mass, receives by far the greater acceleration, that the frame of reference should be determined primarily by the motion of the electron before and after the collision. In particular this is clear if we consider the limiting case in which the velocity of the electron during the impact changes only slightly in magnitude and direction. This offers a possibility of understanding the well-known assymetry in the distribution of continuous X-radiation from an anticathode with respect to a plane normal to the direction of the incident electrons, of which, of course, in its main features the classical theory has given an account. From the point of view of the quantum theory the existence of the assymetry seems to bear strong witness to the fact that processes of radiation can occur in which the electrons after collision with an atom have only partially lost their velocity relative to it*. The paradox, that such an assymetry does not exist for the frequency of the above-mentioned sharp limit of the range of frequency of the radiation, determined by the quantum theory, finds a natural explanation on the basis of our considerations. For it is just a question in this limit of an elementary process, in which the electron is bound after collision and, therefore, on the average possesses no velocity relative to the atoms of the anticathode.

In connection with the question of the frame of reference, the further question arises as to the strict validity of the second postu-

* As Mr S. Rosseland has kindly called to my attention, it seems possible that processes of the nature considered may also play an essential rôle in radio-active transformations. In particular they may give rise to the existence of the so-called continuous β-ray spectra, since the electrons ejected from the nucleus experience great accelerations in the field of force surrounding the nucleus, and might thereby lose any fraction whatever of their energy by processes of radiation. [Translator's Note. Cf. Zs. für Phys. 15, 173 (1923).]

The process of radiation 33

late in the radiative processes considered. Even if, as mentioned, it seems possible to set up a correspondence between the radiative processes and the properties of the motion, which exhibits qualitatively a formal analogy to the relations in multiply periodic systems, the situation is here quite obscure, and the discussion will in much greater measure rest on the assumptions concerning the nature of radiative processes. However, if we follow the analogy mentioned as far as possible, we shall be led altogether to doubt the applicability of the postulate, if we consider such close collisions as those in which, because of the great acceleration of the electron, the radiation-reactions, demanded by the classical theory, would be essential for the description of the motion.

Although we here come upon questions of a completely open nature, still the point of view which we formulated in § 4, in the discussion of the sharpness of spectral lines, seems perhaps to give us further guidance with reference to the limitation and range of the postulates. It also can give us help with respect to the kind of connection which must be sought between the typical applications of these postulates to the problem of explaining the properties of the atoms, and the typical examples of problems of radiation in which the classical theory undoubtedly possesses essential validity, as, for example, the emission of electromagnetic waves in wireless telegraphy. Here we have to deal with a system, the properties of which depend on the collective action of a great number of atomic systems, and in which the relation between the energy and the period, if one can speak at all of stationary states fixed by the conditions (A), would correspond to very large quantum numbers. This circumstance is naturally of essential significance to the problem. It would, however, scarcely be correct to direct the main attention to this side of the problem, and, for instance, regard the applicability of the classical theory in such a case as a direct example of the Correspondence Principle. It is a question here of the applicability of the classical theory to a case far removed from the range of validity of the assumptions which were used in setting up this principle. In discussing this matter, we have, of course, just emphasized that, even in the limiting region of large quantum numbers, the fundamental difference still persists between the ideas of the classical theory, and those which lie at the basis of the applicability of the postulates of the quantum theory. In the problem here considered there is scarcely a question of an asymptotic agreement of the statistical results of the quantum theory with those of the classical theory, but, on the contrary, of a complete breakdown of the postulates of the quantum theory. This is connected with the fact that we have to deal with systems in which the radiation, calculated according to the classical theory, is so great that the energy emitted during a few periods would corre-

B.   3

spond to a great number of elementary processes of radiation of the kind which we meet in the typical applications of the quantum theory to atomic problems. An immediate consequence of this is that the formulation of the postulates of the quantum theory, which was put forward with reference to these typical atomic applications, loses its meaning, as stated, in the case here considered, and that, in particular, the use of the conceptions of the classical theory in applying these postulates is without foundation.

CHAPTER III

ON THE FORMAL NATURE OF THE QUANTUM THEORY

The object of the preceding considerations was to expound the principles which form the basis of the actual applications of the quantum theory to atomic structure, and which will be used as a foundation in the following essays. There still arises, as was mentioned in the introduction, the question of the possibility of forming a consistent picture of phenomena with which these principles can be brought into conformity. Here we have in mind the fundamental difficulties which stand in the way of the effort to reconcile the appearance of discontinuities in atomic processes with the application of the conceptions of classical electrodynamics. Since attention has been directed to various sides of the question, various methods have been proposed for overcoming these difficulties. We shall discuss these methods briefly in the following paragraphs.

§ 1. *The Hypothesis of Light-quanta.* One method of attacking the problem consists in laying weight primarily on the strict retention of such general laws as the conservation of momentum and of energy, even in individual processes. This effort is certainly expressed in the clearest manner in the so-called "hypothesis of light-quanta" of Einstein. As is well known, he has assumed, in order to satisfy the conservation of energy in detail in the processes of emission and absorption of radiation, that even in empty space the processes of radiation cannot, in principle, be described by the use of classical conceptions. According to the hypothesis of light-quanta, the spreading propagation of radiation does not occur in an ordinary wave-motion, but the radiant energy always remains during its propagation in a small spatial region, and is received in the process of absorption as a whole. The quantities of energy contained in these light-quanta are always equal in amount to $h\nu$. Quite apart from the great importance which this point of view has had in placing certain classes of phenomena, such as the photo-electric effect, in a clear light in their relation to the quantum theory, nevertheless, the hypothesis under discussion can in no wise be regarded as a satisfactory solution. As is well known, this hypothesis introduces insuperable difficulties, when applied to the explanation of the phenomena of interference, which constitute our chief means of investigating the nature of radiation*. We can even maintain that the picture, which lies at the foundation of the hypothesis of light-quanta, excludes in principle the possibility of a rational definition of the conception of a frequency ν, which plays a principal part in this theory. The hypothesis of light-quanta, therefore, is not suitable for giving a picture of the processes, in which the whole of the phenomena can be arranged, which are considered in the application of the quantum theory. The satisfactory manner in which the hypothesis reproduces certain aspects of the phenomena is rather suited for supporting the view, which has been advocated from various sides, that, in contrast to the description of natural phenomena in classical physics in which it is always a question only of statistical results of a great number of individual processes, a description of atomic processes in terms of space and time cannot be carried through in a manner free from contradiction by the use of conceptions borrowed from classical electrodynamics, which, up to this time, have been our only means of formulating the principles which form the basis of the actual applications of the quantum theory.

In this connection attention may be called to the attempts, often very ingenious, which have recently been made from various points of view by Whittaker†, to devise a mechanism which reproduces the characteristic features of the quantum theory. These attempts may perhaps point out the direction in which a complete, comprehensive picture of the processes is to be sought in the future. Nevertheless, it must be emphasized that they are scarcely suited, from the nature of the case, to throw light on the actual applications in the present state of the theory.

§ 2. *The Coupling Principle.* Another method of treatment consists in trying to find a universal expression, purely in a formal manner, for the laws of the quantum theory which fix the stationary states of atomic systems, and for those which govern the processes of radiation. This is attained by ignoring at first the propagation of radiation in free space, and considering the field of radiation in an enclosure with reflecting walls. According to classical electro-dynamics, such a field possesses a formal analogy with the motion of a multiply periodic system consisting of massive particles, since the field of radiation, as is well known, can be compounded from purely harmonic, mutually independent, characteristic vibrations. To this case, therefore, the theory for fixing the stationary states

---

* See H. A. Lorentz, *Phys. Zs.* **11**, 349, 1910.
† E. T. Whittaker, *Proc. Roy. Soc. of Edinburgh*, **42**, Part II, p. 129, 1922.

of multiply periodic systems can be applied, at least formally. As is immediately clear, the energy which is thereby associated with each characteristic vibration is equal to a whole multiple of $h\nu$, where ν is the frequency of the corresponding characteristic vibration. This idea forms the basis, as is well known, of the important attempts of Ehrenfest* and Debye† to deduce Planck's law of temperature radiation, without using special assumptions concerning the processes of emission and absorption. The importance of this point of view for our purpose consists in the fact that it permits the condition (B) for the frequency to be regarded formally in the same way as the Conditions of State (A)‡. Since in fact the enclosure, as well as the atomic system, are again in stationary states after the exchange of energy, this exchange of energy can be regarded as governed by exactly the same laws as that in the non-radiating interaction of two atomic systems, which we have discussed in Chapter I, § 4. Since we understand, according to the usual terminology, by the idea of coupling, the possibility of the transference of energy between two systems which are to a first approximation independent, we shall designate such a conception as the "Coupling Principle".

However, in so far as this principle does not establish that only *one* characteristic vibration, with only *one* quantum $h\nu$, can have a share in the exchange of energy between the atom and the enclosure, in contrast to the general case of the interaction of atomic systems, it gives us no direct precept enabling us to account for the requirements of the nature of the emitted radiation which the second postulate of the quantum theory expresses. For the general laws of exchange, however, which are discussed in the preceding chapters, a formulation can be given, in the special case here under discussion of the exchange of energy between an atom and the enclosure, which is conformable to the Coupling Principle, by connecting the conception of coupling with the considerations of probability and with the Correspondence Principle. Just as in classical electrodynamics the so-called force of reaction of the radiation conditions the immediate coupling between the field of radiation and the various harmonic components of the motion of the atom, so we shall assume that the probability of the occurrence of various processes of exchange between the atom and the enclosure is controlled by "latent" reactions of radiation, which answer to the harmonic components corresponding to the respective processes of transition. By taking account of the independence of

* P. Ehrenfest, *Phys. Zs.* **7**, 528, 1906.
† P. Debye, *Ann. d. Phys.* **33**, 1427, 1910.
‡ For more detailed accóunts of the literature see a note of the author's in *Zs. für Phys.* **6**, 1, 1921, where a detailed discussion of the questions concerned is attempted.

the various characteristic vibrations of the radiation in the enclosure, it is, in the first place, reasonable just from the conception of probability that on the exchange of energy between the atom and the radiation the coupling mechanism does not come into action simultaneously for the various characteristic vibrations. In the second place, the conspicuous difference mentioned between the general interaction of atoms and the exchange of energy between an atom and the enclosure, which concerns the limitation of the changes of the quantum numbers in the last-named case, may be brought into relation in this case with the special nature of the coupling. While in the general interaction of two atomic systems, the forces which produce the coupling can be of the same order of magnitude as the forces which act on the particles in the stationary states, the reactions of radiation, which are important for the coupling between the atom and the field of radiation, are always to be considered as vanishing compared with the forces acting on the particles in cases in which a rational definition of the stationary states is at all possible.

These considerations which bring to view a connection of the Coupling Principle with the cycle of ideas of the Correspondence Principle, may at the same time offer a possibility of throwing light on the laws of quantum kinetics for the interaction of atomic systems (see Chapter I, § 4). In this way they may give a point of access for understanding the relations which, notwithstanding the difference in principle between the two cases, exist between the probabilities of the radiative processes of transition in the undisturbed atom, and the probability of non-radiative processes of transition, which can be occasioned by electronic collisions*.

In order to judge the significance of the Coupling Principle it is, however, essential to recognize that its formal applicability is only made possible by the fact that from the first we have ignored the spreading propagation of radiation in free space, in which case exactly, of course, the difficulties in principle of the applicability of the classical conceptions are especially evident. In spite of the formal beauty of the principle, just this limitation makes questionable the significance of the universality obtained by it, in contrast with the dualistic conception which forms the basis of the presentation in the preceding chapter. This presentation is intimately connected with the pronounced dualism already present in the classical theory between the description of the motion of systems consisting of electrified particles, on the one hand, and the spreading propagation of radiant energy in free space, on the other hand. It may, therefore, be more suitable for reproducing certain features which are essential to the actual range of application of the quantum theory.

* See J. Franck, *Zs. für Phys.* 11, 155, 1922.

§ 3. *The Phenomena of Reflection and Dispersion.* As already mentioned in § 1 of this chapter, the consideration of the phenomena of interference is absolutely necessary for the complete presentation of physical facts. Our whole knowledge of the nature of radiation, which to a great extent plays a decisive rôle in the problems of atomic structure, of course rests solely on these phenomena, in the closer consideration of which the formal nature of the quantum theory stands out particularly clearly. This is true not only in the problem mentioned in § 1 of the exchange of energy between atoms and the radiation spreading outwards in empty space, but it must also be emphasized that the explanation of the observable phenomena of interference requires further assumptions which are quite foreign to the postulates of the quantum theory. Thus it seems necessary, in order to account for the phenomena of reflection and dispersion, to assume that an atom reacts on the field of radiation just as a system of electrified particles in the classical theory—in other words, that the atom forms the starting point for a secondary wave-train which stands in a coherent phase-relation with the original field of radiation. To begin with, as already mentioned in Chapter I, it is an immediate consequence of the requirements typical of the quantum theory for the stability of the stationary states, that the reaction of the atom to the influence of a field of radiation in general cannot even approximately be calculated according to the classical theory*. This paradoxical contrast between the classical theory of dispersion and the postulates of the quantum theory is even further sharpened, however, by a closer comparison of the theoretical ideas of atomic structure with the observations. On the one hand, as is well known, the phenomena of dispersion in gases show that the process of dispersion can be described on the basis of a comparison with a system of harmonic oscillators, according to the classical electron theory, with very close approximation if the characteristic frequencies of these oscillators are just equal to the frequencies of the lines of the observed absorption spectrum of the corresponding gas. On the other hand, the frequencies of these absorption lines, according to the postulates of the quantum theory, are not connected in any simple way with the motion of the electrons in the normal state of the atom, since, of course, they are determined according to the condition for the frequency by the difference in energy of the atom in this state and in another (excited) state, essentially different from this one.

---

* This point is especially emphasized by C. W. Oseen (*Phys. Zs.* 16, 395, 1915) in his criticism of Debye's theory of dispersion, in which an attempt is made to calculate the dispersion of gases by means of the classical theory, using molecular models which are based on the quantum theory. See also N. Bohr, *Abhl. über A.*, pp. 138–139, and P. S. Epstein, *Zs. für Phys.* 9, 92, 1922.

According to the form of the quantum theory presented in this work, the phenomena of dispersion must thus be so conceived that the reaction of the atom on being subjected to radiation is closely connected with the unknown mechanism which is answerable for the emission of the radiation on the transition between stationary states. In order to take account of the observations, it must be assumed that this mechanism, which is designated in the preceding paragraph as the coupling mechanism, becomes active when the atom is illuminated in such a way that the total reaction of a number of atoms is the same as that of a number of harmonic oscillators in the classical theory, the frequencies of which are equal to those of the radiation emitted by the atom in the possible processes of transition, and the relative number of which is determined by the probability of occurrence of such processes of transition under the influence of illumination. A train of thought of this kind was first followed out closely in a work by Ladenburg* in which he has tried, in a very interesting and promising manner, to set up a direct connection between the quantities which are important for a quantitative description of the phenomena of dispersion according to the classical theory and the coefficients of probability appearing in the deduction of the law of temperature radiation by Einstein, discussed in Chapter II. It may be of interest to recall in this connection that these coefficients of probability cannot be deduced without further assumption from measurements of absorption spectra. What is directly observed in these spectra is, of course, mainly a weakening of the original wave-train, caused by scattering†. According to the postulates of the quantum theory, the real absorption does not merely consist in a uniform decrease in the energy of the wave-train, but in an exchange of energy, governed by discontinuous laws, between individual atoms and the field of radiation. A proof of such a kind of absorption by means of its effects on the illuminated atoms is obtained in a very instructive manner by observations on the so-called resonance radiation, in which there is no question of coherence with the incident wave-trains‡§.

* R. Ladenburg, *Zs. für Phys.* **4**, 451, 1921.

† As is well known, such a conception has been specially advocated by Julius in connection with his solar theory. In this connection a reference may be made to a recent work by H. Groot, *Physica*, **1**, 7, 1921, in which the analogous problem of the significance of dispersion for the pressure of radiation is treated.

‡ See N. Bohr, *Zs. für Phys.* **2**, 423, 1920, also reprinted in *The Theory of Spectra and Atomic Constitution*, Essay II, where the phenomena of the resonance radiation in their relation to the quantum theory are discussed. See J. Franck, *Zs. für Phys.* **9**, 259, 1922, where the extinction of resonance radiation by the presence of foreign gases as a consequence of collisions of the second kind (Chapter I, Note on p. 11) is discussed in detail.

§ [Note added to the proof]: C. G. Darwin has presented some interesting considerations concerning the interpretation of the phenomena of dispersion, in the quantum theory, in a note which has recently appeared in *Nature* (**110**, 840, 1922). He men-

§ 4. The Conservation of Energy and Momentum in the Quantum Theory.

As a result of the previous considerations, a general description of the phenomena, in which the laws of the conservation of energy and momentum retain in detail their validity in their classical formulation, cannot be carried through. Therefore, we must be prepared for the fact that deductions from these laws will not possess unlimited validity. It is well known that Einstein has not only, as mentioned above, deduced consequences concerning the nature of radiation from the laws of energy, but in connection with his deduction of the law of temperature radiation he has also made use of ideas concerning the applicability of the law of momentum to the radiative processes. He has derived, from considerations of the recoil of the atoms from radiation, arguments for a one-sided, completely directed, emission of radiation. This interesting consideration, which places the incompleteness of our picture of atomic processes in a still sharper light, shows that, in its present formulation, the law of the conservation of momentum, as well as the law of the conservation of energy, do not permit cogent conclusions to be drawn concerning the *nature* of the processes. These laws rather only permit conclusions to be drawn concerning the *occurrence* of those processes which are conceivable according to the postulates of the quantum theory.

As a characteristic example of this type of application of the law of energy, we may consider the assumption that, on a rational definition of the energy of the stationary states, a process of transition connected with the emission of radiation can spontaneously proceed from a certain state only in the direction of states of smaller energy. Although this assumption doubtless can be brought into a certain relation with the Correspondence Principle, yet it is scarcely justifiable to regard it as a consequence of it. On the other hand, the formal character of its deduction from the laws of energy is clearly expressed if we consider the other processes

tions the general failure of the laws of energy in atomic processes and emphasizes the fact that the phenomena of dispersion can be formally explained by the assumption that an illuminated atom acquires a probability of emitting a wave-train, the constitution of which agrees completely with that of the radiation which accompanies a spontaneous transition from a higher stationary state to the normal state of the atom. As Darwin shows, one can in this way attain a statistical agreement with the results of the classical theory of dispersion, if one assumes that this radiation at the beginning of its emission is in a certain phase-relation with the incident radiation. Apart from the fact that the last requirement is scarcely reconcilable with the assumption of the finite duration of the existence of excited atoms, on which the explanation of phenomena of resonance in the quantum theory rests, such a conception encounters apparently insuperable difficulties when it tries to take account of the phenomena of dispersion in very weak illumination. The complete independence of the observed phenomena of dispersion of the intensity of the light (see G. I. Taylor, *Proc. Cam. Phil. Soc.* **15**, 114, 1909; R. Gans and A. P. Miguez, *Ann. d. Phys.* **52**, 291, 1917) might rather call for an explanation of these phenomena in which, as pointed out in the text, an essential feature is a close agreement with the continuous, non-statistical conceptions of the classical theory.

of transition, forced by illumination, assumed by Einstein in the deduction of the laws of temperature radiation, by which the conservation of energy, as defined by means of classical conceptions, seems at once to be excluded.

We meet a corresponding application of the laws of momentum to the processes of radiation in considering the exchange of moment of momentum between the atom and the radiation. The foundation for such a consideration is the assumption, suggested by the cycle of ideas of the Correspondence Principle, that the electromagnetic field of the radiation, emitted in a process of transition, can be compared with a wave-system, such as would be emitted by an electrically charged particle according to the classical theory, which executes pure harmonic vibrations of the corresponding frequency. Such a system of waves possesses a resultant moment of momentum, the ratio of which to the total energy of the waves takes on its greatest value if the orbit of the particle is a circle. In this case it amounts to $1/(2\pi\nu)$, if ν is the frequency of the waves and of the particle. If the total energy emitted amounts to $h\nu$, then the maximum moment of momentum of the field of radiation is equal to $h/2\pi$. If we now consider an atomic system which possesses an axis of symmetry in such a way that the total moment of momentum of the particles remains constant about this axis during the motions in the stationary states, we are led, therefore, on the basis of the conservation of the moment of momentum, to the conclusion that this component of the moment of momentum can never change in a process of transition connected with radiation by more than $h/2\pi$. While such a consideration was advanced by the author* as a support for the consequences drawn from the Correspondence Principle for the possibility of transitions between the stationary states of systems with axial symmetry, it was simultaneously developed by Rubinowicz† independently of this principle. The formal nature of these considerations also, to which Professor Rubinowicz has kindly called my attention in a conversation, appears most clearly in the fact that in the explanation of spectra we must assume the same requirements for the process of absorption in which there can scarcely be any question of a simple conservation of the moment of momentum. We shall return to a detailed comparison of the Correspondence Principle and the law of the conservation of the moment of momentum in individual cases in the following essays.

In connection with the general questions discussed in this chapter it may be pointed out here that in the application of the laws of the conservation of energy and of momentum, such as the

---

* *Q. o. L.* p. 47.
† A. Rubinowicz, *Phys. Zs.* **19**, 441, 465, 1918.

one last mentioned, which is often referred to as a bridge between the classical and the quantum theory, it is rather a question of a formal applicability of these laws to cases in which, according to the nature of the case, the differences in principle of the two theories do not come to light. In this connection the Adiabatic Principle, as well as the Correspondence Principle, occupy a different position, because of their more general range of applicability. They appear, as we shall see, suited, in a higher degree, to point out new ways for further extensions of the quantum theory of atomic structure. As frequently emphasized, these principles, although they are formulated by help of classical conceptions, are to be regarded purely as laws of the quantum theory, which give us, notwithstanding the formal nature of the quantum theory, a hope in the future of a consistent theory, which at the same time reproduces the characteristic features of the quantum theory, important for its applicability, and, nevertheless, can be regarded as a rational generalisation of classical electrodynamics.

COPENHAGEN, UNIVERSITETETS INSTITUT FOR TEORETISK FYSIK.
   Nov. 1922.